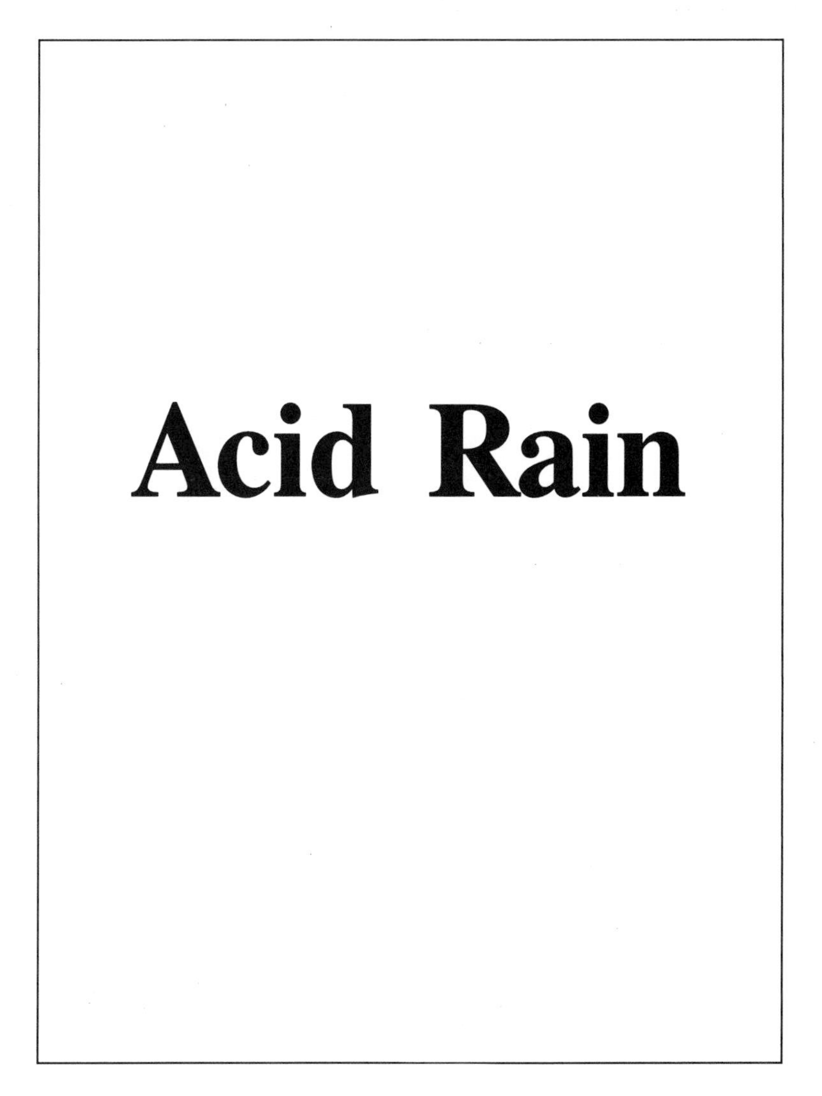
Acid Rain

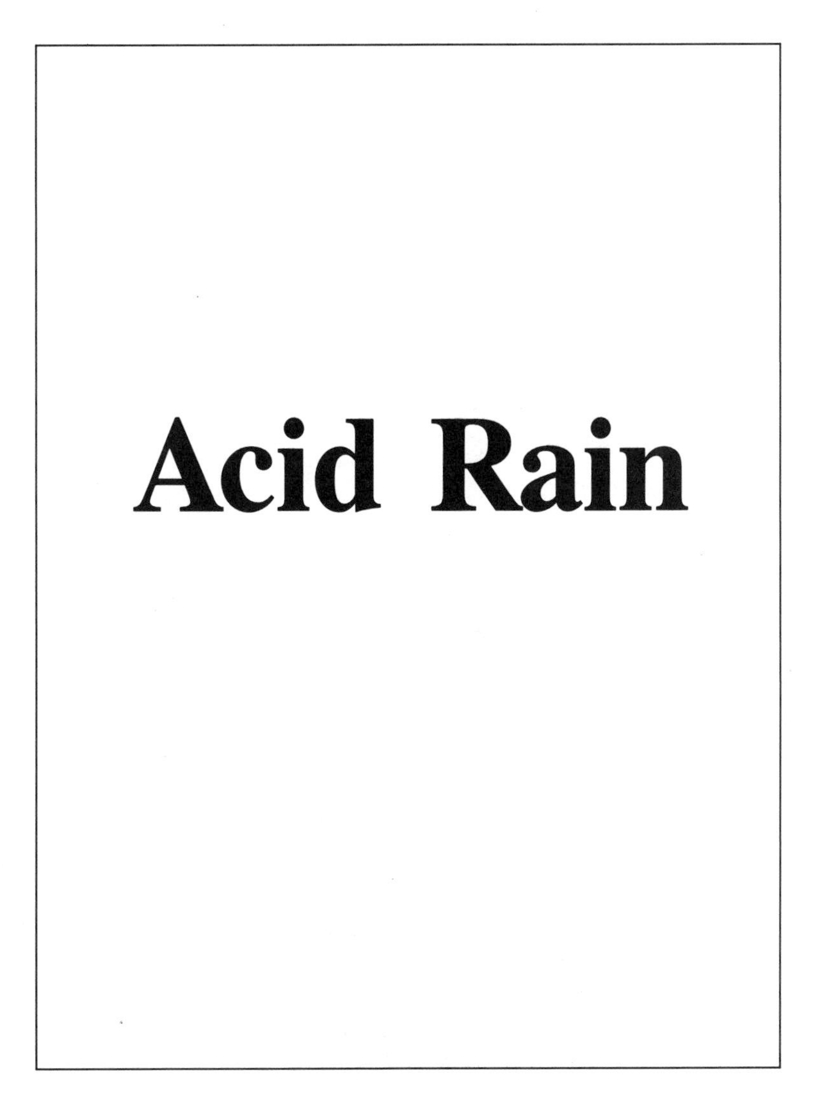
Acid Rain

Look for these and other books in the Lucent Overview series:

Acid Rain
AIDS
Animal Rights
The Beginning of Writing
Dealing with Death
Drug Trafficking
Drugs and Sports
Endangered Species
Energy Alternatives
Garbage
Hazardous Wastes
Homeless Children
The Olympic Games
Smoking
Soviet-American Relations
Special Effects in the Movies
Teen Alcoholism
The UFO Challenge
Vietnam

Acid Rain

by Gail Stewart

LUCENT Overview Series OUR ENDANGERED PLANET

Jav TD195.44.S74 1990t

LUCENT Overview Series OUR ENDANGERED PLANET

Library of Congress Cataloging-in-Publication Data

Stewart, Gail, 1949–
Acid rain.
(Overview series)
Includes bibliographical references (p.
Summary: Discusses the history and causes of acid rain, its effects on the environment and human health,
and what can be done about it.
1. Acid rain—Environmental aspects—Juvenile literature.
2. Environmental health—Juvenile literature.
[1. Acid rain. 2. Environmental health] I. Title.
II. Series: Lucent overview series.
TD195.44.S74 1990 363.73'86 90-5854
ISBN 1-56006-111-1

No part of this book may be reproduced or used in any form or by any means, electrical, mechanical, or otherwise, including, but not limited to, photocopy, recording, or any information storage and retrieval system, without prior written permission from the publisher.

© Copyright 1990 by Lucent Books, Inc.
P.O. Box 289011, San Diego, CA 92128-9011

To My Mother, Barbara

Contents

Introduction

RAIN AND SNOW have always been associated with positive, healthy things. The water that comes from the sky washes the dust and insects from leaves. The water is absorbed into the ground where it is taken in by the roots of trees and flowers. Without rain, crops would wither and die. Rain and snow replenish the lakes and mountain streams. Rain and snow are nature's water supply, and without them all living things would surely die.

Our language, too, is sprinkled with references to the positive associations of rain and snow: Someone who is honest and moral is described as "pure as the driven snow." An idea that seems absolutely correct is said to be "right as rain."

But in the last twenty years, rain has been associated with other, less pleasant ideas. Scientists, especially those who study the environment, have found that not all rain and snow are pure. In whatever form, much of the water that comes from the sky is laced with deadly chemicals that turn the rain to acid. The acid builds up in lakes and rivers. It kills the fish and insects that live in the water. Acid rain has damaged many of our forests. It is harmful to some crops. There is also more and more evidence that acid rain is hazardous to human beings.

Scientists tell us that the deadly chemicals in rain are caused by air pollution. The smoke and exhaust from power plants, factories, and automobiles rise into the air and become part of the atmosphere. This pollution can drift hundreds, even thousands of miles before it comes down in the form of acid rain or snow. Often the place being rained upon is not at all responsible for the pollution in the first place.

There are many differing opinions about the seriousness of the effects of acid rain. Many scientists feel that the problem is so severe that it must be addressed immediately. They say that to wait even two or three years without taking some action would be deadly for life on the planet.

Others, especially many people who work in industry and government, are not as sure about the negative effects. They agree that acid rain does damage to the environment, yet they do not believe that the damage has been shown to be dangerous to humans. They say that it is unfortunate if a few species of fish or trees are killed by acid rain. But, they point out, reducing the air pollution that contributes to acid rain would be a very expensive task. To spend billions of dollars on changing the way power plants and factories work would be foolish until more research has been done, they argue.

The cost of clean air

For that very reason, any serious discussion of acid rain and possible solutions to the problem often turns into a discussion of money. The costs—both of cleaning up the damage acid rain causes and of developing solutions to the problem—are enormous. Since no one wants to bear more than his or her fair share of the financial burden, the result is often finger pointing and assigning blame. Who is responsible? Who will pay? Pollution from Los Angeles's freeways can cause acid rain to fall in Colorado, damaging mountain forests there. Should Los Angeles, then, be made to pay the cost of curbing its air pollution? And if so, who in Los Angeles should pay—the businesses? The taxpayers?

In order to form intelligent opinions about such questions, we need to understand as much as we can about acid rain. Is acid rain a new phenomenon? How does rain become acidic? How are scientists able to measure the kinds of damage acid rain does? Is our planet really in trouble, as some say? Is there something we can do about it before it is too late? By understanding more about the nature of acid rain, we will be better able to deal with the problems it causes.

EXIT 24B
Civic Center
Hill Street
ONLY
110
Downtown
EXIT 24C
Dodger
Stadium
EXIT 1/4 MI

1

What Goes Up Must Come Down

THE TERM *acid rain* is not a scientific one and could be misleading, giving people the impression that rain is the only culprit that brings acid to earth. Actually, the acid in rain showers is only one form of the problem most of us call acid rain. Acid can be found in all forms of precipitation—snow, sleet, hail, mist, fog, and dew. Many of these contain more acid than rain does. In Los Angeles, for instance, scientists have found fog that is hundreds of times more acidic than rain. There are also acids that fall to earth in dry form and are not mixed with any precipitation at all.

Any acid that falls to earth is known by scientists as acid deposition. Deposition means that the material is deposited on the surface of the earth—whether it be lakes, rivers, forests, buildings, or cars. If the deposition is mixed with some form of precipitation, it is called wet deposition. On the other hand, if it comes down in tiny particles of smoke or dust, it is called dry deposition. Whether wet or dry, acid deposition damages the things it touches.

Car exhaust fumes heavily contribute to the dense smog that often hovers above Los Angeles. Cars and trucks are one of the main sources of pollution that causes acid rain.

In his book *Acid Rain*, conservationist Robert H. Boyle states that people's activities have made the sky a sewer. "Each year the global atmosphere is on the receiving end of 20 billion tons of carbon dioxide, 130 million tons of sulfur dioxide, 97 million tons of hydrocarbons, 53 million tons of nitrogen oxides, more than 3 million tons of arsenic, cadmium, lead, mercury, nickel, zinc, and other toxic metals.'' Many of these chemicals are known to be dangerous to humans; some cause cancer, birth defects, or other serious illnesses. Of all of these chemicals polluting our atmosphere, however, only two produce the acid in acid rain—sulfur dioxide and nitrogen oxide.

Sulfur dioxide and nitrogen oxide are released into the atmopshere when fossil fuels are burned. Fossil fuels are the remains of animal and plant life from millions of years ago. The bits of matter have been squeezed together by the shifting and heaving of the earth.

By burning these fossil fuels, people can release the energy that has been stored inside them. Coal, one fossil fuel, is used to run factories and power plants. Natural gas and oil, two other important fossil fuels, are burned as energy for homes and businesses. They are also processed to produce gasoline for automobiles and trucks.

But the burning of fossil fuels has some negative results. As the fuels are burned, they give off poisonous gases. Coal, especially coal that has a high sulfur content, gives off sulfur dioxide. The burning of oil and gas results in nitrogen oxides.

A recipe for acid rain

Sulfur dioxide and nitrogen oxides are the ingredients from which acid rain is made. When they are pumped into the air by a factory's smokestack, for instance, or by the tail pipe of a truck, these ingredients can mix with the moisture in clouds and go through several changes called chemical reactions. Eventually, they become sulfuric acid and nitric acid, the acids present in acid rain.

Scientists know that these changes happen only when the chemicals become airborne. Factors such as water vapor, oxygen, humidity, lack of sunlight, and air current all affect how quickly these chemicals change. So far, scientists are not completely sure why these factors are important. They just know that the change of sulfur dioxide and nitrogen oxides to sulfuric acid and nitric acid takes longer when one or more of these factors are missing.

Once the nitric and sulfuric acids are formed, they may stay in the air anywhere from a few days to a few weeks. How long the acids remain in the air depends a lot on the winds and temperature. While they remain mixed with the clouds, the acids can travel great distances. The acids stay in the clouds until the air gets too heavy and humid. When this happens, water—mixed with acid—falls to the earth as rain.

Some sulfur dioxide and nitrogen oxides fall to earth

In this 1877 lithograph, factories, steamboats, and trains spew grit and soot over the town of Wheeling, West Virginia. As American cities became more industrialized in the nineteenth century, they also became more polluted.

as dry deposition after mingling with certain particles in the air. These particles, often referred to as acid-forming particles, become acidic when they pass through moist air on the way down to earth or when they come into contact with any moisture as they land on buildings, lakes, trees, and other surfaces.

Airmailing pollution

People have known for a long time about the pollution resulting from burning fossil fuels. Cities and towns that had nearby power plants or factories were often grimy and dirty. The pollution from the smokestacks turned many houses and buildings black. The air was gritty, filled with billions of tiny particles of soot. Nobody liked to hang laundry out to dry, for it would become gray. Diseases such as asthma, bronchitis, and pneumonia were far more common in these communities than in places that had no industries.

In 1970 Congress passed a bill called the Clean Air Act. This legislation set strict guidelines for the amount of pollution that utilities and other industries could release into the environment. It was no longer acceptable for factories to blacken the air of nearby communities. Cities and towns close to factories and power plants had to have cleaner air.

One of the ways industries tackled the problem was to build tall stacks. Sometimes known as superstacks, they released the chemical waste high into the sky. It was believed then that pollution thrown high into the atmosphere would somehow be diluted. The concentrations of poisonous chemicals would be lessened because they would be mixed with clean air, or so people thought.

Before 1970 there were only two stacks in the United States that were taller than 500 feet. In the years that followed, however, more than 180 have been built. Some of the stacks are 1,300 feet high, taller than a fifty-story skyscraper.

The results seemed impressive at first. Towns close to large factories and power plants were no longer buried in soot and grime. The tons of chemical emissions were being launched so high that they seemed to be gone forever, scattered in the high winds.

A protest banner from Greenpeace, an international environmental organization, hangs on the smokestack of a copper-smelting company. Greenpeace and other environmentally concerned groups pressure factory owners to stop sending dangerous chemicals into the air.

Long-distance pollution

However, it has become more and more apparent that the pollutants that these stacks pump high into the sky do indeed come down. Even though they no longer fall close to their source, they do eventually fall. Scientists estimate that about half of the sulfur and nitrogen oxides from smokestacks and other sources still trickle down as dry deposition—acid-forming particles of poisonous smoke and dust. The particles usually land within thirty miles of the factory or power station. The rest are "airmailed" high into the atmosphere. Depending on wind speed and currents, the chemicals may travel hundreds, even thousands, of miles.

The effects of pollution are no longer limited to individual communities near utilities or factories. The dioxides travel from city to country, from state to state, even across national borders. What was once a local problem is now a global problem.

Measuring acidity

Acidity, or sourness, is a vague term. However, scientists need to be as precise as possible when talking about the amount of acid in a certain lake, for example. They need to be able to determine whether the acidity has increased, stayed the same, or decreased over a period of time. That is why it is important, when talking about acid rain, to have a way of measuring various degrees of acidity.

The way scientists measure acidity is with a *pH scale*. The letters *pH* stand for "potential for hydrogen." It is helpful to understand what is meant by the various levels of pH.

The pH scale ranges in number from 0 to 14. The numbers close to 0 represent a high degree of acidity. Battery acid, for example, has a pH of 1.8. Lemon juice has a pH of 2.0. Coca-Cola measures a 4.0.

A high number on the pH scale indicates that a substance is alkaline—just the opposite of acidic. Household ammonia is high on the pH scale at 12.0. Laundry detergent measures about 10.0. Baking soda has a pH of 8.2. A reading of 7.0 on the scale is considered neutral—neither alkaline nor acidic. Pure distilled water has a pH of 7.0.

The pH scale is logarithmic and works in multiples of 10. For example, a substance with a pH reading of 6.0 is 10 times more acidic than a substance with a reading of 7.0. A reading of 5.0 is 100 times more acidic than the 7.0. A reading of 4.0 is 1,000 times more acidic than the 7.0, and so on.

What would you guess is the pH of clean rainwater—rain that has never touched a bit of chemical pollution? It may surprise you to learn that unpolluted rainwater has a pH of about 5.6. This high

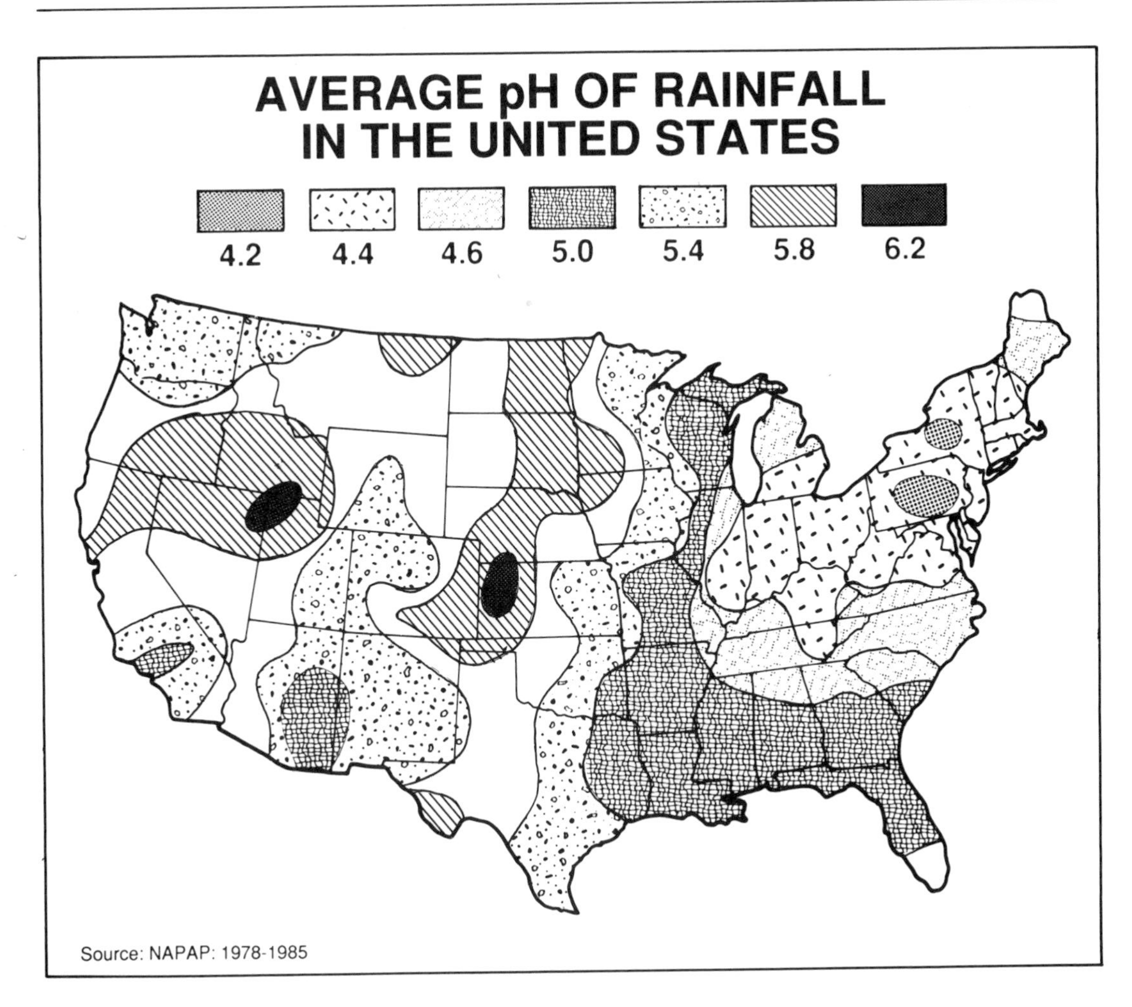

acid level is due to natural causes rather than to people or their factories and cars.

Some of these natural causes include carbon dioxide in the earth's atmosphere, which makes the rain a bit acidic. In addition, sulfur dioxide and nitrogen oxides are released in large quantities when lightning strikes and during forest fires. Even more of the chemicals are spewed out of an erupting volcano. Even the gradual decay of dead plants and animals adds acids to the atmosphere.

But these acids created in nature account for very little of the acid that has been falling to earth. There is nothing harmful about an acid shower of 5.6. Some of the pH levels recorded in the past twenty years, unfortunately, have been many times lower than that.

Rain showers of between 3.0 and 4.0 on the pH scale have been recorded in parts of Pennsylvania, West Virginia, and New York. The lowest reading recorded to this date was in 1978 in Wheeling, West Virginia. There, rain fell that had a pH reading of less than 2.0. That would be close to a shower of lemon juice or battery acid.

Acid levels are increasing in many parts of the world. For instance, many lakes and streams that once had pH readings of 6.0 are now closer to 4.0. Scientific evidence shows that this strongly acidic rain has killed thousands of lakes and forests throughout the

Scientists use this measuring device to gather rainwater and determine its pH level. Rain with a pH of 3.0 or 4.0 is considered unusually acidic.

world. Included are parts of Germany's famous Black Forest, hundreds of lakes in Norway and Sweden, more than fifteen thousand lakes in Canada, and lakes and forests of Adirondack State Park in New York; the list reads like an obituary. And, as scientists warn, the list will keep getting longer. In Canada alone, there are more than 150,000 lakes that measure dangerously high levels of acid.

Many who study the environment say that so much damage has already been done by acid rain that the environment can never recover. They say that if only we had heeded the alarm a generation or two ago, the problem might have been solved. But where was the alarm, and why did no one hear it?

2

The Rain of Years Gone By

ACID RAIN is a popular topic today. Each week, scores of newspapers and magazines publish articles about the effects of acid rain on the environment. However, the subject is not a new one. The term acid rain was first used by a scientist named Robert Angus Smith more than one hundred years ago, in 1872. Yet even before then, people were aware of the kinds of pollution that we now know cause rain to become acid.

Bothering the queen's nose

In thirteenth-century England, people had no idea that burning coal created acid rain. However, many people were aware that this fossil fuel, so common in England, was responsible for a great deal of air pollution. Queen Eleanor, the wife of Henry III, complained in 1257 that when coal was burned to heat buildings and stores, smelly, poisonous fumes produced by sulfur in the coal burned her eyes and nose. Some doctors of the time agreed that the acidlike nature of the smoke was very dangerous to people's health.

In the 1280s the English government established two commissions to control the widespread use of coal as a fuel. King Edward I even made a royal proclamation that anyone caught burning coal would be sentenced to death. Historians report that one man

A cement portion of this bridge is crumbling away. Sulfuric and nitric acids, present in acid rain, are so powerful that they can disintegrate stone and cement.

was indeed executed for burning coal. However, the government soon gave up trying to control its use. It was simply too difficult to regulate a fuel as cheap and available as coal.

As the years went by, more and more coal was burned in England. In fact, coal was really the only fuel readily available to people. Lumber, which was a plentiful fuel in Germany, America, and other countries, was in very short supply in England. By the 1600s the forests of England had shrunk to almost nothing: The lumberjacks who cut down the trees for firewood and lumber had not planted seedlings to replace the cut trees. With no reserves of wood for fuel, the English people had to turn more to their abundant supplies of coal.

"The suburbs of hell"

The air, especially around London and other large cities, became black and gritty with smoke as a result of the burning of coal. A naturalist of the time, John Evelyn, wrote about the thick clouds of coal smoke "which maketh the City of London resemble the suburbs of hell." The air was so polluted with the sulfur of burning coal that street lamps were kept lit even during the day.

British scientist Peter Brimblecombe pointed out that artists of the seventeenth century were well aware of the color of the air. He noted that a sampling of landscape paintings of the fifteenth, sixteenth, and seventeenth centuries shows a change in the color of the English sky. In earlier paintings the sky appears blue, but the later ones show skies of muddy yellows and browns.

In 1620 King James I became concerned about the quality of the air too. He voiced his concern to his aides that the lovely St. Paul's Cathedral in London was crumbling. He was certain that the decay of the building was due to the thick, sulfur smoke caused by burning coal. He did not know what could be done

In an attempt to stop the English people from burning coal in the 1280s, King Edward I proclaimed that the penalty for doing so would be death. Burning coal produced unpleasant and poisonous fumes, but people depended on this method to heat their homes.

about it. Coal was the fuel the country depended on.

The country's reliance on coal increased in the 1700s. Until then, most goods, including clothing and furniture, were made by hand in homes. There were no factories mass producing products on heavy machinery. But the Industrial Revolution in England in the eighteenth century changed the way products were made. Factories were built in London and nearby cities. There, people learned how to weave cloth on spinning machines and large looms run by steam power. The steam was created by burning large amounts of coal.

These factories provided jobs for many workers, so the cities grew larger and larger. The factories, the homes, and the businesses all used coal. The air around London seemed permanently yellow. Doctors reported that there were many times more cases of pneumonia, bronchitis, and asthma in these urban industrial areas than in agricultural areas. Buildings were black with grime. Flowers and bushes could not thrive; most had a greenish gray tinge.

The English government had few alternatives to the pollution problem. To forbid the burning of coal would be a death sentence to the economy. The only solution, which England adopted, seemed to be to build higher chimneys on the factories and to hope that the sooty smoke would be carried a bit farther away from the city.

The forgotten father of acid rain

In the mid-1850s a British chemist named Robert Angus Smith worked for the government as Inspector General of Alkali Works. This post resembled that held by modern air pollution officials. Smith was interested not only in the air but also in the rain in the industrial cities of England. He suspected that an air pollution cloud as thick as the one that hung over London and other industrial cities had to be affecting the

St. Paul's Cathedral in London, like many famous landmarks around the world, is damaged from acid rain. As early as 1620, King James I noticed that the structure was decaying and blamed the damage on the sulfurous smoke created by burning coal.

rainwater passing through it.

For twenty years, Smith collected samples of rain from all over England. He was a precise, careful researcher. So that none of his samples would be tainted with any foreign substances, he used platinum bowls that had been heated until they were red-hot. This ensured that they were free of any outside germs or fingerprints and that the samples of rain in the bowls would be exactly the same as when they came out of the sky.

He published his findings in 1872 in a book entitled *Air and Rain: The Beginnings of a Chemical*

Climatology. In it he revealed that the rain in industrial centers such as Manchester and London was full of sulfuric acid—more so than the rain in nonindustrial areas. He noted that this acid rain, as he called it, could be responsible for metals rusting, buildings decaying, and colors fading on flags and awnings in these cities. He even speculated that the acid rain was causing damage to crops, trees, and plants on the outskirts of cities.

Even though Smith was a respected scientist, his findings were ignored. Modern researchers suggest that perhaps his book was too technical for the average person. A few of Smith's peers might have been interested in his research, but most people were not interested in scholarly reports. The notion of acid rain would lie dormant until the twentieth century.

Filling in the blanks

Many of the experiments done by scientists in the twentieth century actually repeated Smith's research. Two British scientists from London, Charles Crowther and Arthur Ruston, collected rain samples from Leeds, a heavily industrial city. Using the pH scale, they found rain samples of 3.2, which is about 150 times more acidic than normal rain.

Crowther and Ruston, whose field was agricultural science, were interested in how acid rain affects crops. They found that damage from acid rain takes a different form than damage from dry deposition (smoke or soot falling to the ground). The dry deposition, they explained, affects the leaves of the plants, whereas the acid rain sinks into the soil and damages the roots. Indeed, Crowther and Ruston found that acid rain from burning coal had done extensive damage to crops grown near Leeds. Fields of radishes, lettuce, and cabbage had been killed by the rain. Those plants that were not dead were smaller than plants grown away from industrial areas.

In the 1950s two Swedish scientists named Carl Gustav Rossby and Erik Eriksson added to the growing body of knowledge about acid rain. They were interested in the atmosphere—how winds and air currents move pollution from one place to another. They showed that it is possible for all kinds of poisonous chemicals to be absorbed into the air. Depending on the direction of the wind and air currents, the chemicals can be transported hundreds of miles before coming down with the rain or snow.

Factory smokestacks blow chemicals high into the air. There, the wind current can transport these chemicals hundreds of miles before precipitation carries them to earth again in the form of acid rain or snow.

Poison from the cities

In 1955 a Canadian scientist working in England documented the work of Rossby and Eriksson. Dr. Eville Gorham was studying the relationships among plants and animals in mountain lakes, far from industrial cities like London or Leeds. His experiments showed that acid rain falls in these remote lake regions whenever the winds blow from cities. He also observed that the acid rain is capable of upsetting the natural balance of a lake by killing some of the smaller organisms, such as microscopic plants and animals in the water. These tiny organisms provide food for other lake species.

He published his findings, as others before him, in several scientific journals. And, like the others, Gorham found that his work was largely ignored. Gorham had believed that the research would show people how dangerous air pollution and acid rain had become, but no one except the scientific community paid much attention. "There was simply no interest in the subject," Gorham recalled later. "No requests for reprints to speak of."

Scientists today explain the lack of response to Gorham the same way they explain the lack of response to Smith a century earlier. The highly technical scientific journals that Gorham chose to publish his findings in were seen by only a few people. Those scientists who did take an interest in the idea of acid rain did so only from a technical point of view. They were interested in the scientific process of pollution becoming acid, or in the chemical changes of pollutants in the air, or in the patterns of wind and air currents that could carry the acid. These scientists were not looking at acid rain as a problem that had to be recognized and solved.

As Gorham himself later stated, "When I was doing that early acid rain research, I had no environmental consciousness. As far as I was concerned, it was just an interesting situation. And it was my job to put it into the scientific journals. So I pursued it as an intellectual problem, not as a pollution problem. I was simply interested in the question: What determines the chemistry of rain?"

Taking it to the people

A Swedish scientist named Svante Oden is credited with making the general public aware of acid rain. In the 1960s Oden was working to find the source of the acid rain that was falling on Sweden and Norway.

Parts of the outside layer of this column have been eaten away by acid rain. If the damage continues, the structure could be permanently weakened.

Reprinted by permission of UFS, Inc.

Neither of these countries had the kind of coal-burning factories that pump sulfur dioxide into the air, yet a lot of acid rain was falling there.

Oden's research showed that the winds were carrying clouds full of acid from industrial areas in Germany and England into Scandinavia. Oden demonstrated that acid rain was causing damage to wildlife in many lakes and forests in Scandinavia. He was convinced that if the acid rain were to continue, many lakes and forests would be dead in just a few years.

The Swedish experience

Armed with his discoveries, Oden decided to take his research and his predictions directly to the people. He published his scientific findings in the Swedish newspaper *Dagens Nyheter* on October 24, 1967.

The article had immediate results. People who read it were outraged. How was it possible, they asked, that the Swedish people should have their wilderness destroyed by acid rain from other nations? They wondered how many lakes were in danger and how many were dead already. They wanted to know whether the damage could be undone or if it were too late.

Other nations—those that were heavily industrial as well as those that bordered industrial nations—were interested, too. What could happen in Scandinavia could be happening in their countries also. In fact, they realized that some occurrences that no one had had explanations for, such as great numbers of fish dying in various lakes, could be due to acid rain. Those who studied the environment wanted to find answers, and if Oden's predictions were correct, there was no time to waste.

3

The Dying Lakes

AT ABOUT the same time that Svante Oden was publishing his research about acid rain in Sweden in the late 1960s, a Canadian scientist was doing a different kind of experiment. Harold Harvey, a biologist from the University of Toronto, was trying to introduce pink salmon into a lake in Ontario. Harvey had read articles about fishery experts successfully introducing salmon into Lake Superior. The salmon, a popular sporting fish, had done well in Lake Superior, and Harvey hoped to have the same results in Ontario. If his experiments were successful (and there was no reason to think that they would not be), Harvey planned to stock other nearby lakes with the salmon.

An unspoiled lake

The lake Harvey and his team of assistants had chosen was called Lumsden Lake. It was a beautiful, clear, forty-five-acre lake, rimmed by mountains. The water was about sixty feet deep and cold, just right for salmon. Besides its depth and size, Harvey had chosen Lumsden Lake because of its wilderness location. He knew that a lake near a city or town might be polluted from dumps or chemical spills. He knew, too, that choosing a lake near roads or highways could be a mistake, for there would be the danger of exhaust fumes polluting the area. Harvey wanted the lake to be as remote, clean, and unspoiled as possible.

Lakes and streams accumulate acid pollution from falling rain and melting snow. Lakes that appear crystal-clear are often acid-dead lakes; the water is clear because nothing is alive in it.

According to government fish and game statistics,

Lumsden Lake had eight species of fish already, including trout, a fish related to salmon. This was important information to Harvey. Because there were many species of fish that were already thriving, he was sure that the salmon would do well, too.

The case of the missing salmon

Harvey released four thousand tiny salmon, called fingerlings. The following summer, Harvey and his assistants returned to Lumsden Lake to see how the young salmon were doing. They used large nets to scoop the fish out of the water. However, they did not find any salmon. They did not find any trout, perch, or lake herring, either. In fact, the only fish they netted were some white suckers, the fish salmon eat.

Confused, Harvey did a deeper search, going back and forth from one end of the lake to the other. The results were the same. There was not a trace of the salmon. All four thousand were gone, as were the other species of fish that were supposed to be in Lumsden Lake. Only a few white suckers lay in Harvey's net.

There was something unusual about the suckers, too. Many of them were dwarfed. Some had flatter heads than normal, and many had curved backbones. Harvey put tags on the suckers he found before releasing them back into the lake. The following year, he had trouble finding many suckers at all. Of the one hundred suckers he had tagged the previous year, he netted only a few. There were no young ones at all, which indicated that the females were not reproducing.

To try to determine the reason for the disappearance of the fish, Harvey took a water sample from Lumsden Lake to test it for acid. The government data from 1961, ten years earlier, listed the lake's pH level as a healthy 6.8. However, Harvey's water sample showed a pH of 4.4. He checked it again, but the results were the same. There was 100 times more acid in Lumsden Lake than there had been ten years before.

Over the next few years, Harvey and his team of assistants did similar tests on other nearby lakes—150 in all. More than half of the lakes had pH levels below 5.5; many were as acidic as Lumsden Lake. All of the lakes whose pH readings showed a dangerous amount of acid had very low populations of fish.

After years of sampling rainfall in the area, and looking at wind patterns and air currents, Harvey and other scientists found the source of the acid. It was a smelting plant (a factory where nickel is processed) in Sudbury, Ontario, fifty miles to the northeast. The smelting plant was then pumping tons of sulfur dioxide and other poisonous chemicals into the air. Although it was known that the plant had poisoned lakes close by with direct pollution, people were stunned that the poison had actually traveled that far and had done such damage.

Since Harvey's discovery, other scientists in Canada and the United States have been looking carefully at other lakes in North America. Many have been found to be dead—without any fish at all. Thousands more are close to death, with pH levels getting lower (more acidic) every year.

A quiet death

Harvey's research was the first real study showing the death of a lake. His findings reveal that there are very gradual changes that come about before a lake loses its fish. In fact, the death of a lake does not begin with the death of its fish. According to Harvey's research, the changes begin with the microscopic plant and animal life in a lake. The changes are so subtle, so quiet that they may not be noticed at first. When a lake begins to die, there are usually no piles of rotting fish on the shore, nothing that would alert people to the danger. "There is no muss, there is no fuss, there is no smell," Harvey reports. "The fish quietly go extinct."

To understand how acid can kill a lake, it is

necessary to understand what an ecosystem is. An ecosystem is the balance of life within a certain territory. That territory could be very small, like a pond, or large, like a huge forest or lake. Even planet earth is part of a highly complex ecosystem.

All the living things in the territory of an ecosystem rely on each other. In a lake, insects, fish, turtles, waterfowl, frogs, and maybe water mammals, such as muskrats or beavers, depend on each other to exist. These animals also depend upon plants that grow in and around the lake. There are also tiny organisms, too small to be seen except with a microscope. These organisms are valuable to the lake's ecosystem, for they are responsible for the decay of plants and animals. When a fish dies, or when a leaf falls in the water, these organisms quickly break down the material into tiny pieces. These pieces provide food for the tiny fish and minnows in the lake.

Each species of life is important in the ecosystem. If even one part of the system were removed, the whole system could become weak and even die. For instance, if those tiny organisms that help the decay process were removed, there would be no food for the small fish. Many of the small fish would starve. Larger fish, turtles, loons, and other species that feed on those small fish would suffer, too, since their supply of food would be scarce.

Many kinds of acid death

Harold Harvey and other scientists have discovered that some species of fish and other water animals are more vulnerable than others to acid. When the pH of a lake falls to 4.5, the lake is considered dead—no living thing can survive in the water. Yet there are many species that die long before the pH falls that low.

At a pH of 5.5, for instance, clams, snails, and crayfish disappear. Their shells, which are their protection from predators, dissolve in the acid. Also at

The life and death of a lake ecosystem

To support normal aquatic life, a lake habitat must have a balanced food chain and allow for normal reproductive cycles. When acid is introduced to the ecosystem, either through acidic rainfall or runoff from acidic snowpack, the influx of acid may alter the lake's overall pH, creating a more acidic environment. Also, acidic runoff can release heavy metals—highly toxic for many aquatic species—from rocks and soil. Sensitive organisms may be wiped out and resistant life forms are weakened as their food supplies are affected.

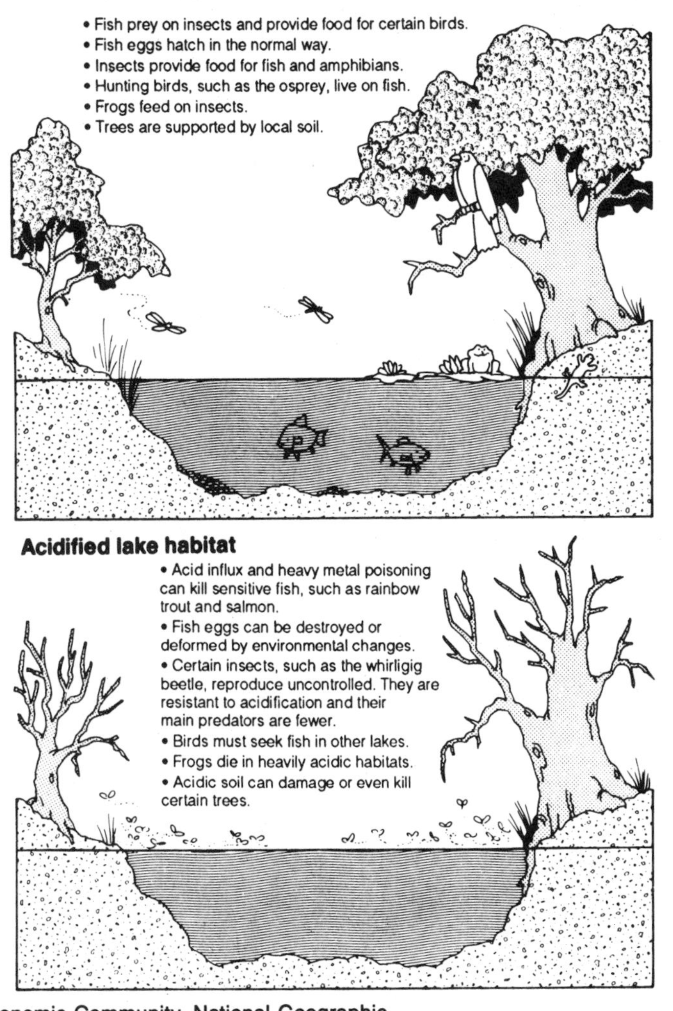

Data/European Economic Community, National Geographic
Source/InfoGraphics

risk are salamanders, tadpoles, bass, trout, and salmon.

Scientists have discovered, too, that acid rain affects the young fish more than the adults. Sometimes the newly hatched fish are deformed or too weak to survive. The acid in the water robs the fish of the calcium they need for their blood and bones. Often the hatchlings are born with twisted, oddly shaped backbones—a sign that they lacked calcium. In other cases, there is so little calcium that the fish eggs do not have outer shells or the shells are too thin. When this happens, the young never hatch.

Big fish are a bad sign

Dr. Carl Schofield of Cornell University has done much research on the effects of acid rain on fish and other water life. He points out that one of the clues that a lake is dying is that there are no young fish. "In a normal lake you would expect to find predominantly younger fish," Schofield says. However, when a lake becomes acidified, the population soon switches to older, larger fish.

Schofield says this explains why people who fish on dying lakes often catch the biggest fish. There are no tiny fish for the bigger ones to feed on, so they begin to kill and eat one another, leaving only the largest fish. Once those big fish are caught or die,

This healthy trout lived in a lake that was free of acid pollution. The fish has a well-formed body of normal size.

there are no young fish growing up to replace them. Soon, there are no fish at all in the lake.

There is another way that acid rain can kill fish in lakes. Acid rain can mix with soil on the shore, enabling certain dangerous ingredients of the soil to wash into the water. This can cause fish to actually drown.

Soil contains traces of many metals: mercury, copper, lead, zinc, and the most common metal on earth, aluminum. These metals are not harmful as long as they are bonded, or stuck, to the soil. Aluminum, in fact, is a kind of "glue" in nature that bonds metals so that they are harmless. However, the acid in acid rain can break the bonds that hold metals to the soil. When the bonds are broken, the metals become very dangerous to living things.

The aluminum danger

It does not take a great deal of activated aluminum to be dangerous. Scientists estimate that one tablespoonful in a lake the size of a football field would kill all the fish. Aluminum causes an irritation to form on the gills. As the irritation increases, the fish produces a white, gooey substance, called mucus, to cool the burning sensation. The burning does not go away, however, and the fish continues to produce mucus. Eventually, the mucus clogs the fish's gills, and it suffocates, sinking to the lake bottom.

This trout from an acidic lake clearly shows the harmful effects of acid rain. The body is deformed and thin. Often, acid in water prevents young fish from getting the calcium they need for proper bone formation.

Mercury, another metal released by acid rain, does not kill the fish. However, mercury is poisonous to humans. The mercury runs off the soil into the lake, just as aluminum does. It is absorbed by the fish and collects in the muscles and tissues. People who catch and eat fish containing mercury can become very sick and die. Mercury poisoning is a threat in many places around the world. Many lakes have signs warning people who fish about the possibility of fish being contaminated with mercury. The signs warn people not to eat too many fish caught in the lake.

The danger of mercury to human beings was first observed in 1953 in Japan. A plastics factory had been dumping mercury into the sea, and it entered the bodies of many fish. Although the fish did not die from the mercury, the metal poisoned the people who ate the fish. Forty-four of the people died immediately, and many others were left paralyzed.

Acid shock

Acid rain falling on a healthy lake does change the pH of the lake, but only after many years. The acid becomes diluted by the lake water, so the lake water does not become acidic right away. But there is a phenomenon that does cause a dramatic, almost immediate change in the acidity. This phenomenon takes place in the spring, when nearby snow and ice melt and run off into the lake.

Scientists call it acid shock because the lake's water becomes highly acidic in a short time. The snow and ice, which are heavily polluted, collect throughout the winter. In some cases, the acid snow is as acidic as vinegar.

When the temperature rises in the spring, the ice breaks up and the snow melts. The melted snow, heavily laden with acid, rushes into the lake. For several days, the lake has more acid than a rainfall could cause, especially in the shallow water near shore. After the spring floods are over, the lake's acid

level may become more balanced as the acidic water becomes diluted in the deeper waters of the lake, but unfortunately, the damage has already occurred.

Acid shock occurs in the spring, when acid snow and ice from the mountains melt and flow into lakes. Because the snow accumulates acid pollution all winter long, it dramatically raises the acidity of a lake when it melts.

The damage that acid shock does to fish is severe. Ross Howard, a science writer for a Canadian newspaper, calls acid shock "the cruelest stage of acidification—it hits the lakes when their aquatic dwellers are most vulnerable, and with deadly results." The fish are vulnerable because the spring thaw is the time of breeding and egg laying for many fish. Most often, the breeding and egg laying occur in the shallow, acidic water. Some of the adults survive, although the eggs are usually destroyed by the acid.

A lake whose plants and animals have died because of high acidity is called acid-dead. An acid-dead lake is surprisingly beautiful. The water is crystal clear and reflects the brilliant blue color of the sky. People who hear of a lake being acid-dead are often astonished that it looks as clean as it does. The clean, clear water is a sign that the lake is missing the tiny bacteria and other organisms that break down dead plants and animals. A healthy lake's water is full of these bacteria; they give the water a cloudy, murky appearance.

A leaf falling into a healthy lake will be broken down by these organisms. Nutrients and energy will be released, and there will be more food for the algae and small fish. A leaf falling into a dead lake, on the other hand, will remain there. Eventually, the leaf will sink to the bottom. Without the bacteria to rid the lake water of such debris, the bottom of the lake will, over time, become thick with leaves, branches, and dead plants and animals.

Acid rain researchers in Sweden have identified another sign of an acid-dead lake. A slimy mat of algae, known as sphagnum moss, invades the lake bottom. Ecologist Robert H. Boyle wrote that the moss is "thick enough to be picked up and shaken like a rug." Sphagnum moss, which usually grows only on land, is not damaged by acid rain—in fact, it appears to thrive on the acidic water. As it lies on the lake bottom, it suffocates any plants that might still be trying to grow. Bacteria that do not need oxygen survive under the carpet of moss. They produce foul-smelling gases that occasionally come bubbling up to the surface of the lake.

Lakes at risk

Not every lake is in danger of dying because of acid rain. Even in areas that receive very acidic rain and snow, there are some lakes that are not immediately at risk. Yet scientists are usually able to predict what lakes are in danger.

There are some substances in nature that buffer, or weaken, acids. Limestone is a rock that contains some of these substances. Lakes that rest on limestone are not as vulnerable to acid death as other lakes. There are chemicals circulating in the lake water of limestone lakes that buffer the acid rain.

On the other hand, lakes that rest on granite or other hard rock are vulnerable to acid death. Granite has no buffering chemicals within it. Unfortunately, there are many such lakes in the United States and Canada—more than 1.5 million square miles of North America rests on such rock. The lakes in the Boundary Waters Area of Northeastern Minnesota fall into this category. So do all the lakes in the Appalachian Mountains of the eastern United States and the Rocky Mountain lakes. Many thousands of lakes in Canada also rest on a hard granite base. These are the areas at risk, and scientists are monitoring the pH levels of these lakes carefully.

4

The Forests in Decline

SCIENTISTS ARE generally in agreement about the damage acid rain does to lakes and streams. But there is some disagreement about the ways acid rain is damaging ecosystems located on land, such as forests. In the last few years there has been much discussion in the media about trees mysteriously dying in many of the world's forests. The famous Black Forest in Germany, the Green Mountains of Vermont, and the forests of the southern Appalachians in North Carolina have all been losing trees rapidly. Many scientists feel that acid rain could be responsible.

Waldsterben

It would be difficult to find a nation anywhere that takes as much pride in its forests as Germany does. The German people have a strong bond with their Black Forest which covers more than seven thousand square miles. The forest is important for the lumber it produces—over one billion dollars' worth every year. It is a scenic setting and one that has historic importance for Germany. In 9 A.D., German tribes defeated the army of the Roman Emperor Augustus in the Black Forest. The forest is also the setting for many old German fairy tales, such as the stories of Hansel and Gretel and Little Red Riding Hood.

Mt. Mitchell in North Carolina is an area of forest decline. The trees there are dying as a result of their exposure to acid fog.

With so much national pride in the Black Forest,

it is no wonder that the people of Germany are horrified at what has happened to it in the past decade. Tall spruce trees stand like skeletons. A few have yellow, dry needles; most have no needles at all. The trunks of the trees are smaller than they should be. The boughs droop, and twigs hang straight down from the branches. Foresters call this the "silver tinsel effect," because the twigs look like Christmas icicles.

When the trees began looking sick in 1976, most foresters in Germany were not alarmed. There had been a drought that year, and it was not unusual to lose a few trees. But by 1979 there was cause for alarm. There had been plenty of rainfall, yet many trees looked as though they were not getting water. Older evergreens were not the only trees that became sick. Younger trees, too, were losing needles.

The mysterious death

The rapid decline of the Black Forest was mysterious. Drought had definitely been ruled out as a cause. So had insects and all known tree diseases. Scientists examined some of the sick trees and were puzzled. The German foresters named this mysterious, deadly decline of the Black Forest *Waldsterben*, meaning "forest death."

In 1982, six years after it was first noticed, *Waldsterben* had spread to the large oak and beech trees. Many of these trees were hundreds of years old and had survived storms, droughts, and other hazards.

Today, visitors to the Black Forest are appalled at the destruction. One writer noted that the place was "like a scene from a doomsday movie." Gerrit Mueller, a forester from Germany, is depressed by what he sees. "It is frightening," he agrees. "Some people think that in five years there will be no forest left. To think that an entire forest might die is incredible."

Waldsterben is not limited to Germany, either. Forests in Czechoslovakia, Austria, and Poland have

Chancellor Helmut Kohl of West Germany examines the damage to trees in his country's famous Black Forest. Since 1976, trees have been dying in this forest, and many scientists speculate that acid rain is the primary cause.

all suffered declines like the one in the Black Forest. In North America, too, there have been major declines in forests in North Carolina, Vermont, and Quebec. Many scientists in Europe, Canada, and the United States are working hard to find the cause of the declines. Although there are clear signs that acid rain may be playing a part in the problem, the answers are far more complicated.

A complex ecosystem

The death of a forest appears to be a far more difficult mystery to solve than the death of a lake. One reason is that trees live so much longer than fish. Acid rain damage to a tree with a life span of three centuries might not show up for thirty or forty years. On the other hand, a newly hatched fish in water with a low pH will show signs of sickness in a week or two.

Also, it is difficult to duplicate forest conditions in a laboratory. Scientists studying the effects of acid rain on lakes can easily hatch trout in acidic water and watch the results. However, too many of the conditions that affect trees are impossible to manufacture—insects, cold winters, pollution, elevation, and abrupt changes in rainfall. All of these conditions put stress on trees in a forest. The occurrence of unexplained illness in a forest, or forest decline, may be a combination of one or more of these stresses, although fingers are pointing more and more to air pollution in the form of acid rain.

Hubert Vogelmann, a botanist from the University of Vermont, has been studying the decline of a forest called Camel's Hump in the Green Mountains of Vermont. He says that over the past twenty years, the forest has changed drastically:

> Twenty years ago the evergreen forests on the slopes of Camel's Hump . . . were deep green and dense. The red spruces and balsam firs that dominated the vegetation near the mountaintop thrived. . . . The trees were luxuriant, the forest was fragrant.

Dead spruce trees appear next to healthy ones on White Face Mountain in New York. In addition to acid rain, other stress factors, such as drought and disease, can destroy a forest. Scientists often have difficulty determining the specific damage caused by acid rain.

> Today the red spruces are dead or dying and some firs look sick. Gray skeletons of trees, their branches devoid of needles, are everywhere in the forest. Trees young and old are dead. . . . The brittle treetops often break off, leaving only a jagged lower trunk with a few scraggly branches. . . . The forest is collapsing. It looks like somebody dropped a bomb up there.

Vogelmann and other scientists have worked hard trying to understand the cause of the decline of the forest at Camel's Hump. Today, Vogelmann is convinced that air pollution, especially in the form of acid rain, is largely responsible for the death of Camel's Hump. He is also convinced that because of the complexity of a forest ecosystem, such a theory can never be proven.

Vogelmann compares himself and his fellow researchers to medical researchers. Although most medical researchers are convinced that cigarette smoking causes lung cancer, they also know that it can never be proven. Some smokers never get lung cancer. Some nonsmokers get lung cancer. How is it possible to say that in a complex machine like the human body, smoking is *the* cause of the disease?

"The complexity of the natural environment is so great," Vogelmann says, "that it would be difficult to say that beyond doubt acid rain is responsible for everything we've observed. But there is a lot of experimental evidence that we have that points in that direction. A lot of it."

These bare trees died when their leaves were damaged and no longer able to absorb sunlight. Acid pollution destroys leaves by burning through the waxy, protective coating and harming delicate cells.

What the evidence shows

Trees are like filters in the air. They screen all kinds of particles passing through the air around them. These particles collect on the leaves of the tree. Current research has shown that when these particles contain acid—either in dry or wet form—they can damage a tree's leaves. Because the leaves are the part of a tree that makes its food, any damage to them is harmful to the health of the entire tree.

Leaves have a thin protective coating. This coating feels waxy to the touch. Scientists know that acid rain can burn through this protective layer, leaving brown spots on the leaves. Especially delicate are the tiny guard cells on the surface of each leaf. The guard cells control the opening and closing of microscopic pores on the leaf that let moisture and sunlight in. Acid rain can damage the guard cells, leaving them unable to send the signals to the pores to open or close. When this happens, the leaf pores may remain open, "drowning" the leaf in too much water; or the pores may remain closed, suffocating the leaf by cutting off its supply of light and air.

Evergreens at risk

Pine trees are also vulnerable to damage from acid rain. A pine tree's needles are constructed to nourish the tree after the needles fall to the ground. On each needle are whole colonies of tiny bacteria and algae that help the tree change nitrogen into food at the roots. These bacteria can be burned away by acid rain. This leaves the tree without an adequate food supply, thereby making it weaker and more vulnerable to disease.

The more time a tree spends with acid rain on its leaves, the more damage it will undergo. Scientists have observed that the hardest-hit trees in forests tend to be the ones on the tops of mountains or hills. These are the trees that spend much of the day surrounded by low clouds or fog. The trees on Camel's Hump, for instance, get more than half of their moisture from fog.

Mt. Mitchell, an area of forest decline in North Carolina, is also covered in fog most of the time. In fact, out of every ten days, Mt. Mitchell's trees spend eight bathed in fog. Scientists there have found that the fog has a very low pH (high acidity). On its least acidic days, the fog has a pH of 2.9, more acidic than vinegar. The lowest pH recorded at Mt. Mitchell was

2.1, which is about the acidity of bottled lemon juice.

Robert Bruck, a botanist who has been studying the forest decline at Mt. Mitchell, has found more and more evidence that points to acid rain as a cause. He and his researchers set up a miniature greenhouse on Mt. Mitchell. The greenhouse had two chambers. On one side were young trees that had a supply of the same fog and air as the rest of the Mt. Mitchell forest. The second chamber had a filter on its fog and air supply. The filter removed sulfuric and nitric acids, key ingredients in acid rain. After six weeks in the greenhouse, the trees receiving the unfiltered air drooped and were losing leaves. The other trees, those whose air was filtered, were thriving. They showed no damage to leaves whatsoever.

A scientist from the U.S. Department of Agriculture tests the effects of acid rain on trees. By misting trees with acidic solutions, he can observe how they react.

Harming the forest floor

After acid rain does its damage to the needles or leaves of the tree, it ends up on the forest floor. The acid can damage the tree in a number of ways once it reaches the ground. The amount of damage depends on the nature of the forest's soil.

Just as lake water or rainfall can be measured for pH, soil also has a level of acid. As with lakes, soil that contains limestone has low acidity and is able to withstand acid rain. The limestone can buffer the acid so that it does not damage the trees.

Acid in the soil can do damage to the forest by releasing aluminum, which is normally not harmful. However, when activated by the acid, the aluminum becomes highly poisonous to trees. The aluminum enters the tree's tiny, hairlike roots, choking them. When these roots become clogged, water from the soil cannot get to the upper branches of the tree. Slowly, the trees die of thirst, even though there may be plenty of moisture in the soil. The occurrence of aluminum poisoning seems to coincide with the recent declines of forests, such as the Black Forest. Scientists have found that the level of activated aluminum in forest soil has tripled since the 1960s.

Besides activating aluminum in the soil, acid rain kills important organisms on the forest floor. Any forest floor is covered with fallen twigs, animal and bird droppings, and dead leaves. Many types of bacteria and fungi feed on these materials. This process is called decomposition, or decay. The decaying material is broken down into tiny pieces that work their way into the soil. Under the soil, earthworms speed up the process of decomposition by breaking the decaying matter into tinier pieces that can then supply nourishment to trees and other plants.

Acid rain is harmful to the decomposition process. Acid kills many of the bacteria and fungi that live on the forest floor. When the pH level of soil drops to 4.0, earthworms die. Without the earthworms and the bacteria to decompose the debris, the material builds up on the forest floor. In forests that are suffering declines, such as the Black Forest and Camel's Hump, debris is twice that of a normal, healthy forest. In some places, the dead leaves and droppings are almost two feet high.

When debris on the forest floor builds up like this, tiny seedlings from the tree may not be able to survive. They are not able to work their way down to the soil to take root. Because of this, the forest cannot rejuvenate itself. There will be no young trees to take the place of older trees when they die.

A cycle of catastrophe

Once large trees in a forest begin to die, other parts of the forest ecosystem are in danger, too. Large trees provide an "umbrella" of shade in the forest. Without this umbrella, more sunlight reaches the forest floor. Plants that thrive in sunlight compete with young trees for growing space. On Camel's Hump a recent survey found not one healthy, thriving seedling on the forest floor.

In the winter months, the absence of large trees

results in more snow collecting on the forest floor. The deep snow makes it hard for deer and other nonhibernating animals to find food. Bitter winter winds can penetrate a forest that has lost its biggest trees. Animals and birds that cannot keep warm die in great numbers.

Scientists know that this cycle can result any time a forest suffers a major decline—whether the cause is acid rain, drought, or disease. Although many scientists are convinced that acid rain is a major cause of many of the world's forest declines, other scientists feel that more research is needed to solve the puzzle of *Waldsterben*. As more is understood about acid rain and complex forest ecosystems, more pieces of the puzzle will perhaps fall into place.

5

Acid Rain and Human Health

MOST PEOPLE live miles from a forest or lake. These people may not be as concerned about the effects of acid rain as people who do live near forests and lakes. The destruction of a lake's ecosystem may be unfortunate, they say, but not risky for them.

These people may be surprised to learn that the effects of acid rain are not limited to forests and lakes and the life that exists in them. A growing amount of evidence indicates that acid rain is dangerous to people, harming the body in a number of ways. As one medical researcher said, "This is not just about a few species of salamanders or pine trees. This hits closer to home. It is about how easily you breathe, or how clean the water you drink really is. Is it hard to work up emotion about a salamander dying? Then think about your health, the health of your family and friends."

In December 1952, a dense blanket of fog sat over London. It stayed longer than most episodes of fog, for there was a layer of moist, heavy air that weighed it down. Under normal weather conditions, air pollution around London rises from smokestacks and is blown away by the winds. This time, however, the pollution was trapped at ground level by the fog. More than four thousand people—most of them young children or people with breathing problems—died

A policeman stands guard in a thick fog that surrounds London in 1955. Many people died in London in the 1950s when fog trapped polluted air over the city.

from breathing the polluted air. Many thousands more suffered damage to their hearts and lungs.

Much of the poison in that air was made up of sulfur dioxide and nitrogen oxides, the main ingredients of acid rain. Scientists know now that the fog of 1952 was an acid fog. There have been other instances of acid fog trapped by moist air. In the Meuse Valley in Germany a similar fog killed sixty people in 1930. Several hundred more suffered lung and heart damage. In Donora, Pennsylvania, an acid fog in 1948 lasted two days. Eighteen of the small town's residents were killed by the acid pollution.

Episodes such as these are rare. Most of the time pollution is not trapped close to the ground, and people are not forced to breathe such poisonous air. However, cases such as these have made some scientists wonder about the long-term effects of these poisons. Even though "killer fogs" may not be smothering people and striking them dead, are people being harmed over a longer period of time in equally dangerous ways by acid pollution? Many scientists say yes. There is an increasing amount of evidence that acid rain may be doing damage to our bodies. Such damage may be more sinister, they say, because it often does not show up for years.

Avoiding the body's defense systems

One of the threats to the human body comes from the poisons that cause acid rain. Before the sulfur dioxide and the nitrogen oxides go through chemical reactions to become acid rain, they are tiny particles in the air. They may come from the exhaust of a bus or car or from the smokestack of a huge power plant. Whatever the source, they may be carried for long distances by winds. The particles, when inhaled, can be dangerous for people with breathing problems, such as asthma or bronchitis.

The human body has defenses against particles that could be harmful if inhaled. The hairs in the nasal passages trap many of the larger particles, such as dust or pollen. Mucus in the nose also traps the particles, keeping them from reaching the lungs.

There are times when people do not breathe through their noses at all. Those who are exercising or doing strenuous work need air more quickly and breathe through the mouth. The body has defenses for mouth-breathers, too. Tiny hairs called cilia wave gently back and forth in the lungs, carrying out unwanted particles.

Unfortunately, the particles that cause acid rain are far too tiny to be trapped by mucus, nasal hair, or cilia. These poisons can get into the body through the nose and through the mouth. They are so small, in fact, that they are drawn very deep into the lungs. The delicate lung tissue becomes burned and irritated.

People living near smokestacks that emit sulfur dioxide and nitrogen oxides often have respiratory problems. These poisons enter the body through the nose or mouth and then burn the lung tissue, making breathing difficult.

"GEE, SORTA TAKES YOUR BREATH AWAY, DOESN'T IT?"

In a healthy person with no breathing problems, the particles can cause a bad cough, bronchitis, or a chest cold. In a young child, an old person, or someone with a breathing problem, the reaction can be much more severe. The irritated lung tissue can set off a chain of reactions that make the muscles around the lungs get tight. People who suffer a reaction to sulfur dioxide and nitrogen oxides often feel they cannot get enough air, no matter how deeply they breathe.

Choking on pollution

If the episode is very serious, the person's lips and fingernails look blue, indicating that his or her blood does not have enough oxygen. Breathing becomes more difficult, and the heart has to work much harder than normal. In such cases, the person might have a heart attack or slowly choke because of a lack of air.

The idea of such a death may seem farfetched. However, according to the Congressional Office of Technology Assessment, more than 51,000 Americans and Canadians die this way every year. The cause of these deaths is inhaling sulfur pollution according to that office.

A study done at the Brookhaven National

Laboratory in Upton, New York, in 1982 indicated that sulfur pollution is a widespread health danger. That study estimated that between 5 and 8 percent of all deaths in the United States during that year were caused by inhaling these chemicals.

Children need clean air

Yet most scientists are aware that the studies are not conclusive. Just as it is difficult to prove that forest decline is brought about by acid rain, the causes of human death and illness are hard to pinpoint. There are many factors that enter into a person's ill health. Perhaps there are hereditary factors that make that person more likely to suffer breathing problems. Perhaps there are factors such as diet or allergies that bring on severe problems.

Milton Anderson is a New Jersey doctor who treats children with breathing problems. He feels that acid pollutants cause a great deal of respiratory illness. He also admits that his theory is almost impossible to prove:

> I think it's gotten to the point where doctors know that whenever the sulfur pollution gets really bad, we're going to see a lot of very sick kids. I don't need anyone to prove that to me. There don't have to be dead bodies in the street for us to understand the threat of acid pollution.
>
> I think pretty soon most people will realize that it's not just a coincidence that folks who live in high-sulfate areas get lots more breathing problems. There may be all sorts of variables that make that hard to prove. But there's just too much evidence to ignore.

Some of that evidence came from a study of 1,400 Canadian children aged seven through twelve. The children lived in one of two small towns—Tillsonburg, Ontario, or Portage la Prairie, Manitoba. Tillsonburg has a high level of acid air pollution. On the other hand, Portage la Prairie has exceptionally clean air. Researchers wanted to know whether there was any difference in the children's health in the two towns.

As it turned out, there was a significant difference. The children from Ontario had far more chest colds, stuffy noses, allergies, and coughs than the Manitoba children. In addition, the Ontario children did not do as well on a test to measure the working ability of their lungs. Scientists suggested that the irritated lung tissue of these children caused them to become out of breath more easily.

A serious health threat

Another study, which also points to the conclusion that acid pollution is a health threat, is one by Guy Orcutt of Yale and Robert Mendelsohn of the University of Washington. Orcutt and Mendelsohn studied over two million death certificates from 1980. They also studied weather maps and other reports that gave each day's sulfate pollution levels. From these data, the two researchers matched up days when the pollution was quite severe in certain parts of the country with deaths from heart and breathing problems. They concluded that 187,686 deaths in the United States that year were caused by acid pollution.

"We really should be thankful we're up here above the acid rain."

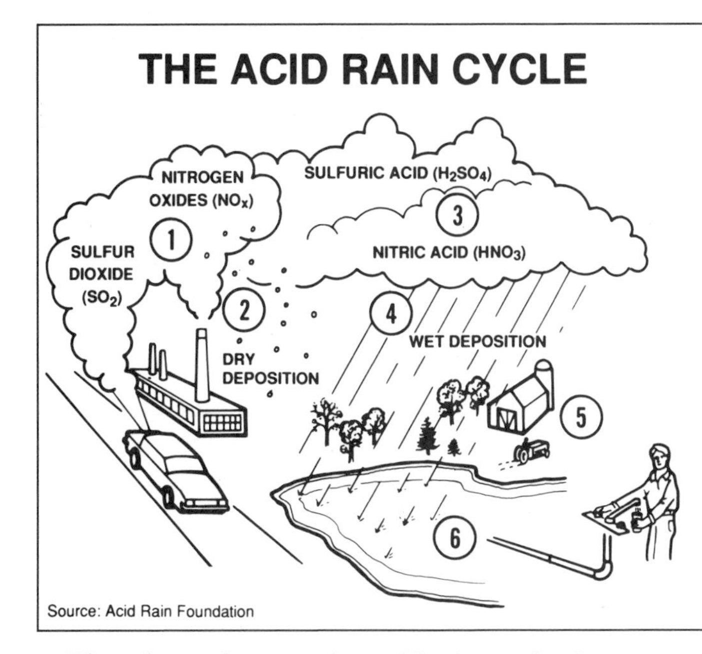

Source: Acid Rain Foundation

(1) Exhaust from automobiles, power plants, and factory smokestacks fills the air with sulfur dioxide and nitrogen oxide gases. (2) Some of the gases become attached to particles in the air and fall to the earth as dry deposition. (3) Great quantities of sulfur dioxide and nitrogen oxides mix with the moisture in clouds to form sulfuric acid and nitric acid. (4) These acids fall, with the rest of the water in the clouds, as acid rain, or wet deposition. (5) The acids poison trees, crops, and other plants. (6) Sulfuric and nitric acids build up in rivers and lakes, killing fish and polluting the water.

There is another way that acid rain can be dangerous to humans, a way that is far easier to prove. It is known that the earth contains many metals that could be dangerous to humans—lead, mercury and aluminum, for example. Most of the time, these materials are harmless because they are bonded to other elements in the soil.

However, when acid causes these metals to become unglued from surrounding rocks or soil, they can be carried deep into the ground. From there, they may find their way into underground streams. Many of these are the sources of our water supplies.

Medical researchers know that these metals can be dangerous, even fatal, to human beings. Aluminum, for instance, can kill people who have kidney problems. Aluminum can also collect in the brain tissue of healthy people. Some scientists suspect that aluminum deposits in the brain are the cause of Alzheimer's disease, a condition that results in memory loss, nervous system problems, and death.

Some areas are more susceptible than others to acid rain damage. As is true with lake water that rests on limestone, ground that is made up of limestone can buffer the acid in polluted rain, so that it cannot do as much damage. When the acid is weakened, it is not able to unglue the harmful metals from the soil or rock. On the other hand, areas composed of granite or other hard-rock soils, such as in the northeastern United States and Canada, are not able to buffer acid rain.

So far, the amounts of metal in drinking water have been very small—not nearly as much as in the contaminated fish. However, the fact that the metals are there at all gives scientists reason for concern. Because metals do not leave the body once they enter it, little amounts may accumulate into large amounts over a lifetime. No one yet knows how large an amount will prove to be harmful to people.

Pipe poisoning

Acid rain can damage drinking water in other ways. The acids can loosen bits of metal water pipes. Although modern plumbing is often made of plastic, older homes have copper pipes. The copper pipes are joined together by a mixture of lead and tin. Lead is extremely dangerous to humans—even small doses can cause problems in the brain and nervous system.

One study done in Ontario, Canada, found that water that had been sitting in plumbing pipes for ten days had very high levels of copper and lead. This standing water had more metal in it than did the source of the water supply. This could be a widespread danger, since there are many people who do not shut plumbing off when they go on vacation. Any water left standing in the pipes would certainly be dangerous to drink. For that reason, health agencies urge Canadians to let the water run for a few minutes before drinking it.

Acid rain can also dissolve the reinforcements around large water pipes. In some parts of the United States, such as the San Francisco Bay Area, asbestos is used to reinforce the cement around water pipes. Asbestos is not harmful when it is bound to cement; however, it is highly dangerous when it becomes loosened from cement. It has been linked to cancer and other serious diseases. Health officials worry that when acid rain comes in contact with the cement, loose asbestos could find its way into a city's water supply.

6

Acid and the Built Environment

THERE IS no doubt that acid rain is a threat to the different segments of the natural environment, including humans. Scientists are also concerned about acid rain's damage to what they call the "built environment." The built environment consists of objects people have created.

Some of these objects are practical; they make life easier, safer, or more comfortable. Houses, for example, keep people out of rain and snow and safe from intruders. Automobiles, airplanes, and bridges are built to enable people to move about in the world quickly and safely. All of these objects are important to people's comfort, and they are being damaged by acid rain.

Irreplaceable objects

There are other objects in our built environment that are not practical but are extremely valuable. Historic landmarks and statues, old cathedrals and temples, paintings and sculpture—these are parts of the built environment that cannot be replaced. As environmental author Thomas Pawlick writes, they are "a part of our national soul."

Acid rain damages these and other objects of the built environment. Just as airplanes, bridges, and cars are being eaten away by acid rain, treasures of art and

This statue, which stands outside the Field Museum of Natural History in Chicago, is being defaced by acid rain. The black parts of the statue's nose and chin have had greater exposure to acid showers than its cheeks and eyes.

history are being destroyed.

Whether in dry or wet form, acid pollution can damage the surface of objects made of wood, stone, brick, concrete, metal, or fabric. (The acid can eat right through nylon—especially the nylon used to make umbrellas.)

Automobiles are susceptible to acid damage. The Environmental Protection Agency (EPA) estimated recently that half of all rust on automobiles is caused by acid rain, acid slush, or acid snow. A car's paint, too, is affected by acid rain. The acids cause the colors of the paint to change. In many cases, the paint blisters and flakes off.

Small raindrop-sized blotches appear on cars after an acid rain shower. In one case in central Pennsylvania, hundreds of cars were scarred by these polka dots after a rain shower with a pH level of 2.3. More and more signs have appeared in gas stations and car washes advising people to wash the acid rain dust off their cars.

Bridges, bricks, and B-52s

Many factors determine how much damage acid rain can do to the built environment. The amount of rain, the nearness to power plants or factories, the direction of the wind, and the humidity all determine the amount of decay and corrosion from acid rain. An area that has a large amount of fog or humidity tends to suffer more damage than dry areas.

For that reason, many steel bridges located over water get rusty and corroded by acid. When metal is corroded, it becomes weaker and cannot take the stress of weight. Acid sulfur pollution was blamed for the 1967 collapse of a bridge between West Virginia and Ohio. The accident resulted in the deaths of forty-six people.

The railroad industry is concerned about acid rain, too. The steel track must be checked often for signs of rust and corrosion. In England, one-third of the

This bridge over the Ohio River collapsed in 1967, killing forty-six people. Acid rain supposedly corroded the steel bridge and made it too weak to support traffic.

money spent in replacing track is the result of acid rain corrosion. In addition to causing delays and route changes, replacing track costs millions of dollars.

The metal in airplanes is also susceptible to damage from acid. The United States Air Force spends more than $1 billion every year to repair or replace corroded airplane parts. Robert Summitt is an engineering consultant to the Air Force. He has seen the corrosion on the B-52 bomber, and he is surprised at how fast the damage sets in. He said that rust in some airplanes is "so bad that you can pop the rivets out of the wings with your fingers."

A Swedish study in 1986 found that metals rust four times faster in areas that receive a lot of acid rain. Yet the damage is not limited to steel and other metals. Houses and buildings made of brick and stone crumble more quickly in acid rain areas, too. Acid pollution dissolves the mortar, which is the cement used to hold bricks together. When the mortar wears away, the bricks crumble more easily. They shift and cannot bear the heavy weight of the bricks above them.

Scarring the face of years gone by

As expensive as bridges and automobiles are, they can be replaced. Many of the objects acid rain is destroying, however, are works of art—pieces of sculpture and architecture that cannot be replaced.

Many of the bronze statues in town squares throughout the country are pitted and stained. Some of the features of the statues are becoming distorted and blurred because the acid has worn them away. Of course, some erosion over the years is normal. Wind and weather can smooth and wear away even the rough surfaces of mountains. Yet scientists are certain that much of the damage is not erosion. They can see telltale signs of acid rain.

It is especially easy to see acid damage to statues of people with horses. Rain collects in the little hollow places behind a horse's ears. Because the moisture stays in those spots longer, the acid has more time to eat away at the metal. Soon the hollow places turn into holes. The state of Massachusetts spends $8 million every year to repair acid damage to its many statues.

Statues made of bronze and copper naturally turn green after a while. The green covering, called a patina, is a protective shield that keeps the metal from being damaged by wind and water. However, acid pollution destroys the patina. The acid dissolves the green covering, leaving in its place a streaky, black coat.

The Statue of Liberty, America's most famous landmark, has received a lot of acid damage over the past two or three decades. Miss Liberty underwent extensive repair in the late 1980s. During that repair, workers noticed that there were black holes and thin spots on the green patina. Officials of the renovation project wondered whether the worn areas were the result of years of winds and salt spray from the harbor or if acid rain had done damage. Experts were called in, among them Robert Baboian, a specialist in the process of acid corrosion. He collected tiny scrapings from different parts of the statue and analyzed them thoroughly.

Baboian found that the worn spots were from acid rain as well as exposure to the salt water in New York Harbor. The acid was making the statue's outer layer quite thin in some sections. The thin, blackened parts were, Baboian suggested, far more susceptible to corrosion and rust than the rest of the statue. The National Park Service, whose job it is to maintain the landmark, knows that extreme care must be taken with Miss Liberty. Plans to wash the statue with a mild detergent were abandoned after the acid tests—the patina is far too delicate.

The U.S. Capitol building in Washington, D.C., has also suffered acid damage. The dome, which is made of white marble, has craters all over it. The tiny craters are about one-quarter of an inch in diameter—raindrop size. Dr. Erhard Winkler, a geology consultant to the government, studied the dome carefully. "It looks like shrapnel has hit it," he observed.

The treasures of Europe

The damage to historic landmarks and buildings is not limited to the United States. Just as countries in Europe are finding that their lakes and forests are in danger from acid rain, they are also discovering that national treasures are being damaged.

There are, for example, more than 100,000 large stained-glass windows throughout Europe. Many of them are several hundred years old; the oldest was created one thousand years ago. The beautiful windows, with panels of brilliantly colored glass, have

Experts discovered in the 1980s that acid rain was destroying the outside layer of the Statue of Liberty. Because this layer is now thin and weakened, the entire statue is more vulnerable to the damaging effects of wind, rain, and pollution.

The white marble dome of the U.S. Capitol building in Washington, D.C., has tiny craters covering its surface. Acid rain has eaten through the stone and produced these craters.

survived centuries of battles, earthquakes, and storms. Many of them were taken down temporarily during World War II and stored so that they would not be shattered during the bombings there. Yet those windows, which survived so much, are in danger of being destroyed by acid pollution. The acid harms the windows by wearing away the color in the glass until it fades away altogether. Some experts warn that unless there is a big decrease in the acid pollution Western Europe receives, the windows will be completely faded in another ten or twenty years.

Dr. T. N. Skoulikidis, a specialist in acid corrosion, is worried about the fate of some of the beautiful buildings in his homeland of Greece. Some of the endangered buildings, such as the famous Parthenon, have stood for more than two thousand years. Acid rain, much of it carried in the clouds from the industrial cities in Poland and Germany, is wearing away some of the detail on the buildings. The stone is turning to a fine powder, and the buildings are crumbling. Skoulikidis has looked at photographs of the buildings taken throughout the twentieth century. Based on these, as well as other historical records, he does not believe that the damage can be blamed on natural erosion. In fact, he reports that more damage has been done to the ancient buildings in the last twenty years than in the previous 2,400.

The same claims can be made for another well-known monument. Although it is not as old as the ancient Greek structures, the famous Taj Mahal of India has suffered much acid damage in the past two decades. The Taj Mahal was built between 1623 and 1645 by an Indian ruler as a tomb for his wife. Its lovely towers, decorated walls, and domes have attracted many visitors for centuries.

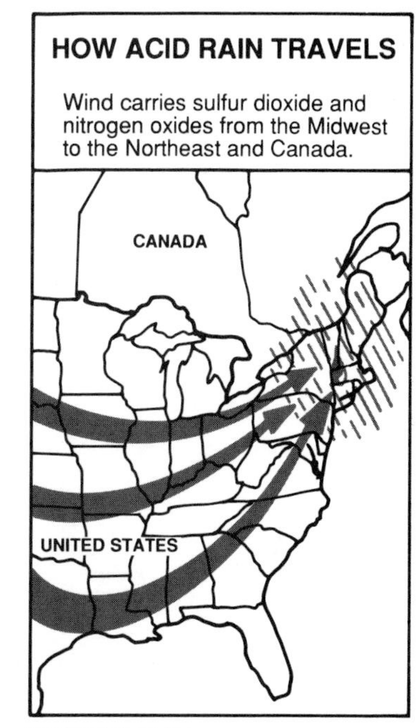

Source: *Midwest Living*

India has never been an industrial center like the United States or Germany. However, in 1970 an oil refinery was built upwind of the Taj Mahal. Acid pollution, in both wet and dry form, is eating away at the delicate lines and features of the monument. There are workers whose full-time job it is to monitor the building's damage. They constantly dust off the gritty particles of pollution so that the acid does not stay long on the marble surfaces.

Acid damage inside

Exteriors are not the only parts of buildings to experience acid damage. Interiors of buildings also suffer, even though they are not directly exposed to acid rain. Tiny particles of sulfur dioxide and nitrogen oxides can enter buildings through heating and air-conditioning systems. The particles can circulate inside the buildings and fall onto books, paintings, wall hangings, and rugs.

Some directors of art museums and libraries have noticed the harm acid pollution has done. Leather-

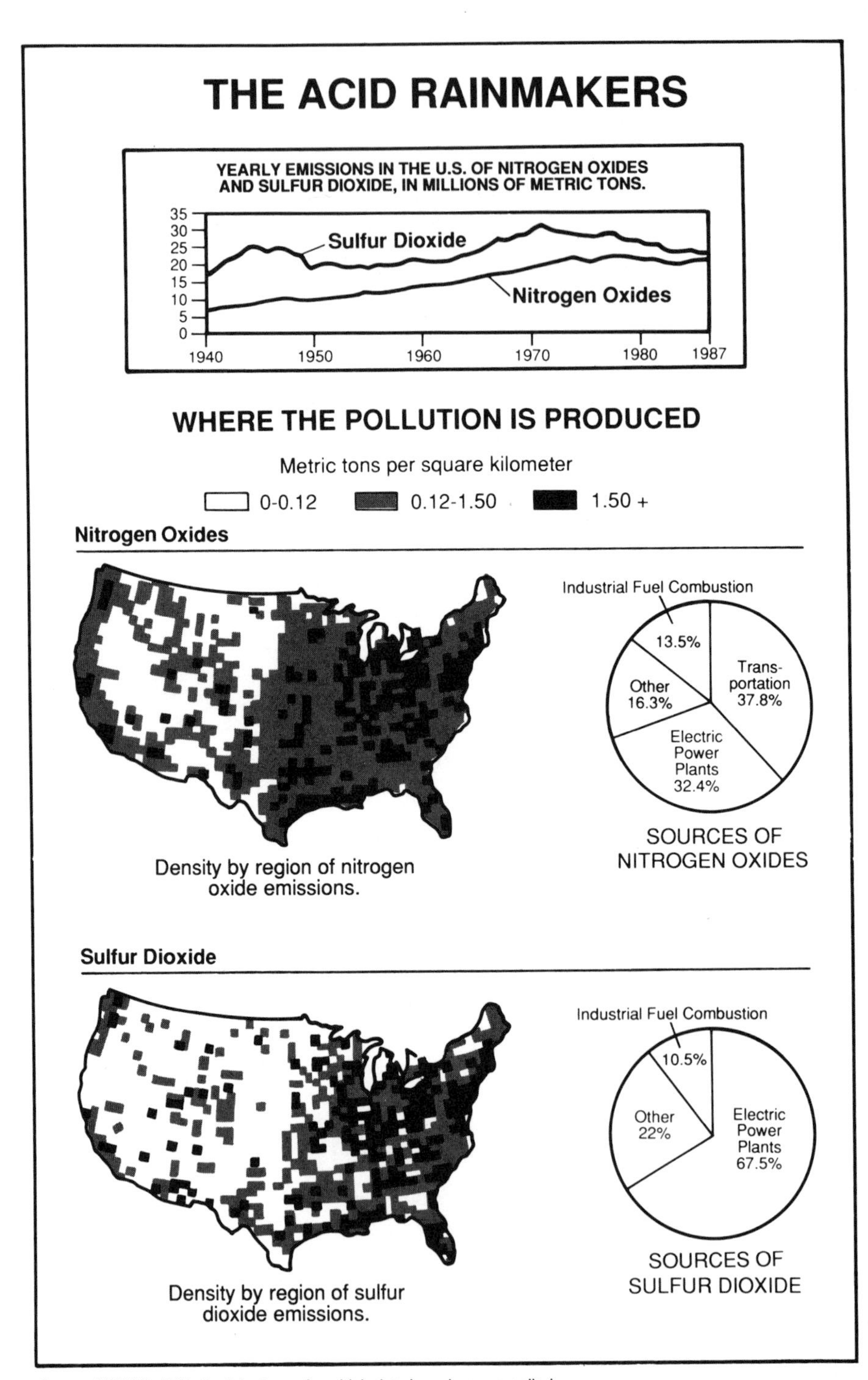

Source: NAPAP: 1980, the latest year for which data have been compiled.

bound books are susceptible to damage. So are oil paintings, microfilm, tapestries, and historic documents. All deteriorate more quickly if acid pollution comes in contact with them.

Many of the world's most treasured art and documents are kept in museums that are located in heavy industrial areas. Chicago, New York, London, Paris all receive a great deal of dry acid pollution from nearby factories and power plants. Because of the fear of losing their treasures, some museums have installed special filters, called scrubbers. The scrubbers, which cost millions of dollars, trap the harmful acid particles before they can circulate indoors. The Library of Congress in Washington, D.C., and Chicago's Newbury Library are two buildings that have installed such scrubbers.

BayWa

7

Finding a Solution

FEW PEOPLE doubt that acid rain is a problem. However, in the United States there has been a great deal of disagreement on how serious the problem is and how it should be solved. Organizations such as the National Wildlife Federation, the American Lung Association, Greenpeace, and the Sierra Club insist that acid rain is a very immediate problem. They say that the United States has the technology today to reduce the pollution and that it is just a matter of spending the money to do it. Unless the industrial countries of the world stop polluting very soon, many of the world's lakes, forests, and wildlife will die, according to these organizations.

On the other hand, there are other groups—especially those representing American power companies, the oil and coal industries, and car manufacturers—who are not convinced. They know that it would cost them billions of dollars to reduce the pollution they pump into the air. They feel that there should be more studies and research before they are asked to make any drastic changes in the way they produce their products.

A lime solution is sprayed over the forest floor in Germany to counteract the acid rain that accumulates in the soil. Lime neutralizes acid and raises the pH level of the soil.

Not a local issue

The problem is even more complex, however, because a lot of pollution from U.S. factories and power plants does not stay in the United States. Acid pollution often travels hundreds of miles before coming down as acid rain. In the case of American-made acid rain, much of it heads north, crossing the border into Canada.

Canadian leaders are anxious for the United States to control its pollution. They point to studies showing that between 50 and 75 percent of Canada's acid rain comes from American power plants. Already more than fourteen thousand Canadian lakes have been poisoned by acid. Many thousands more are on the brink of being poisoned. Canada is a nation that relies on its tourist industry. Poor fishing and acid dead lakes would not attract campers and people who fish. Millions of tourist dollars would be lost.

Pollution problems in Canada

For the last ten years, Canada has tried to work with U.S. leaders in devising a plan to eliminate acid rain. In 1980 the two nations signed a Memorandum of Intent, which was an agreement that both countries would make acid rain control a priority. They promised to work hard at coming up with ideas to cut down the amount of sulfur dioxide and nitrogen oxides they pump into the air.

However, Canada has accused the United States of moving too slowly. While Canada has pledged to cut its sulfur dioxide emissions in half over the next ten years, the United States has made no such pledge. On a number of occasions, President Reagan did not send an American representative to important international acid rain conferences. Canadian leaders, such as Prime Minister Brian Mulroney, have stated that America does not seem as committed to a solution to the acid rain problem as it should be. The acid rain issue has caused strained relations between the two countries.

However, President Bush has promised to make acid rain legislation a priority. He has promised to work more closely with the Canadian government to solve the problems caused by acid rain.

For the United States, the heart of the problem is money. Certainly the power plant officials and the coal and automobile company presidents do not want dirty air or dead lakes and forests. However, they do want proof that spending billions of dollars is absolutely necessary. They do not want to cut their profits by spending money on ideas that are not guaranteed to work. Scientists and researchers maintain that, even though absolute proof is impossible, enough evidence that acid rain causes environmental damage has been gathered. They say that the time has come for action, not more studies.

In 1988, Canadian prime minister Brian Mulroney (center) addressed the U.S. Congress to ask that the United States reduce its factory emissions. Mulroney claimed that these emissions contributed to the acid rain that was killing many of his country's lakes. Former vice-president George Bush (left) and former house speaker Jim Wright (right) are also pictured.

Dealing with the damage

What kinds of action have been proposed? According to environmentalist William Watson, the ideas for dealing with acid rain range from "the ridiculous to the absolutely ingenious."

Some of the ideas are temporary, "quick-fix" solutions, designed to keep a bad situation from growing worse. A process called liming is one such solution. By adding lime to acid-damaged lakes and streams, scientists can buffer the acid in the water. Huge quantities of lime are poured into the water from boats or dropped from airplanes. Scientists have tried liming lakes in Canada, the United States, Sweden, and Norway. In most cases, the lime reduced the acid in the water and raised its pH level.

However, liming a lake does not restore it to the way it was before the acid damage. If acid has killed off bacteria, minnows, or trout, liming the water will not bring those species back. Also, aluminum that has been unglued by acid rain from nearby soil and rocks and has washed into the lake will still kill fish.

Opponents to liming lakes say that the process only works for two or three years. After that, the lime is dissolved by the acid, and more lime must be added to the water. Liming a lake is expensive, too. Estimates range from $250 to $350 per acre of water. Since there

COST OF DEACIDIFYING LAKES			
		Annual Cost of Liming	
REGION	ACRES OF ACID LAKE	BY HELICOPTER	BY BOAT
Adirondacks	4,846	$1,211,500	$242,300
Southern New England	5,669	1,417,200	283,400
Central New England	480	120,000	24,000
TOTAL NORTHEAST	12,496	3,124,000	625,000
Upper Midwest	2,628	657,000	131,000
TOTAL IMPACT AREA	**15,124**	**$3,781,000**	**$756,000**

Source: NAPAP Interim Assessment 1987 Living Lakes Data.

are thousands and thousands of lakes threatened by acid damage, the cost could be billions of dollars.

Scientists insist that liming is only a Band-Aid solution to a large problem. According to acid rain expert Professor Gene Likens of Cornell University, liming is not an answer to acid rain because it is directed at the symptoms, not the cause, of the problem. "[Liming] can be compared to taking morphine before you cut your leg off," Likens explained. "It might ease the pain, but you still bleed to death."

Cleaning up "King Coal"

Rather than try healing the damage acid rain has already done, scientists agree that it makes more sense to prevent acid pollution. More than two-thirds of the acid pollution in the United States originates from power plants. If sulfur dioxide and nitrogen oxides could be eliminated before they were released into the air, there would be no damage to worry about later.

Power plants exist to create electricity for our homes, schools, and businesses. The process is complicated and involves the use of steam power to turn magnets around coils of wire. This produces electricity. Steam comes from hot water. It is the way the water is heated—by using coal—that produces the pollution that causes acid rain.

For centuries coal has been the fuel used by factories, homes, schools, and businesses. In the United States coal has always been plentiful and for that reason has been a cheap source of energy. Because of abundant supplies and the many people who depend on the production of coal for their jobs, it has often been called "King Coal."

Each power plant in the United States uses millions of tons of coal every year. In fact, one large power plant in Ohio burns more than six hundred tons of coal every hour to produce steam. Although the burning of coal does result in a great deal of sulfur dioxide, it is possible to remove the poisons so that they are not released into the atmosphere.

Washing, scrubbing, and eating coal

Scientists have found that coal that has been "washed" burns more cleanly. The coal is crushed into smaller pieces, then sprayed with jets of water. After being passed through screens, the coal is gradually separated from pieces of dirt and rock, impurities that add to the pollution when burned. Finally, the coal is spun very fast in a machine. The spinning motion forces the impurities to the bottom of the machine. One advantage of washing coal is that it becomes lighter. If it is washed at the site where it is mined, the coal can be transported more cheaply.

Another process, called scrubbing, can remove almost 95 percent of the sulfur dioxide after the coal is burned. (While the previously mentioned scrubbers installed in some libraries and museums clean air as it comes in from the outside, these large industrial coal scrubbers clean the gases from the coal fires before they go out the chimney of a power plant.) In the scrubber, the hot, poisonous gases are sprayed with a mixture of water and lime. The sulfur dioxide mixes with the lime and water to form a gray, gooey substance called sludge. The sludge falls to the bottom of the scrubber.

All new power plants are equipped with scrubbers.

Older plants that are required by law to remove the sulfur dioxide from their emissions, however, must pay to have a scrubber installed. At a price of $150 million each, they are not cheap. Many power companies complain that the scrubbers are expensive to operate, also. They claim that more than one-third of the plant's budget is needed to run and maintain a scrubber. At some of the larger plants, a staff of sixty workers is needed to keep the scrubber running.

Another problem with scrubbers is the sludge itself. Thus far, no one has been able to come up with a use for it. It does not burn, so it cannot be used as fuel. It is useless as an insulator or building material. At present, the sludge from scrubbers is dried and hauled to landfills so that it does not pollute the ground or water supplies.

One of the most fascinating ideas for reducing the amount of sulfur dioxide produced by burning coal

A helicopter drops lime over an acid lake to temporarily weaken the acid. Eventually, more acid rain will dilute the concentration of the lime solution and return the lake to dangerously acidic conditions.

was developed by a researcher from the University of Minnesota. Microbiologist Henry Tsuchiya discovered a bacterium that actually eats the sulfur on raw coal. (Sulfur, when burned in a power plant, becomes the acid rain-producing sulfur dioxide.) Tsuchiya found that by turning the bacteria loose on water-soaked coal, the tiny organisms would eat 98 percent of the sulfur in less than two weeks. The idea is still in its beginning stages, but fellow researchers are excited by the possibilities of Tsuchiya's hungry bacteria.

Politics and coal

Most of the coal that is used in American power plants today is high in sulfur. It is sometimes called "dirty coal," since it gives off lots of soot and dust when burned, due to its high sulfur content.

High-sulfur coal is mined in Ohio, Illinois, Virginia, and West Virginia. Many hundreds of thousands of people are employed in the high-sulfur coal industry. Some are miners; others work in railroad transportation or are involved in processing the coal.

Some types of coal, on the other hand, are naturally low in sulfur. Such coal would not need the same amount of washing or scrubbing as high-sulfur coal. It might seem to be an easy answer—switching from high-sulfur coal to low-sulfur coal. Certainly, it would be better for the environment, since less sulfur dioxide would be released into the air.

The "easy answer," however, is not really easy at all. Low-sulfur coal is found in the western states. It would cost more for the coal companies to ship the coal to the busy industrial centers of the East and Midwest. Even more threatening, it would mean the loss of thousands of jobs in states that rely heavily on the coal industry. Entire communities have sprung up around coal processing plants in the East and Midwest. These communities would suffer economic decline if the mines were shut down.

This wind turbine in Texas generates electrical power without polluting the environment. By developing new sources of energy that do not burn fossil fuels, governments and industry can help eliminate the causes of acid rain.

It is easy to see why congressmen and other representatives from the states in the Midwest that produce high-sufur coal fight the switch to low-sulfur coal. They represent coal workers and their families who want to keep their jobs. To support the switch, which would be good for the environment, would be extremely unpopular with the voters. And, since popularity is what keeps politicians in office, acid-rain prevention has taken a backseat to employment in these states.

Prevention, not cure

Doctors and other health professionals urge people to stop smoking, to exercise, and to eat healthy foods. They believe that prevention is as important as curing disease. Many environmentalists believe the same to be true about acid rain. They say that rather than be concerned with liming lakes or otherwise cleaning up the damage acid rain causes, people should try to prevent the acids from getting into the environment. By reducing the amounts of sulfur dioxide and nitrogen oxides we pump into the atmosphere, we can cut down the amount of acid rain.

The problem, these environmentalists feel, is that our world relies too heavily on fossil fuels. Over the coming years, people must learn to use energy more wisely. New sources of energy, such as solar and wind power, need to be developed. Technologies that clean coal are helpful for now, but what happens in the long run depends on wise decisions about the types and amount of fuels we use.

Scientists agree that our planet does not have the luxury of many years to delay. Thoughtful use of the environment must begin today. Strong leadership from the president and members of Congress is important, especially in adopting legislation against the kinds of pollution that cause acid rain.

Education and cooperation

However, the real strength of any program will depend on how informed we all are. By reading and listening to all the information we can find about acid rain, we can become more knowledgeable about its effects on planet earth. Writing letters to the president, representatives, or congressional leaders is a good way to let them know how strongly we feel about acid rain. Because they are elected representatives, they appreciate knowing how voters (and future voters) stand on important issues.

Solving the problem of acid rain requires the cooperation and clear thinking of all of us. The health and well-being of our planet depend on our good judgment.

Glossary

acid deposition: Sulfuric acid or nitric acid falling to earth in either dry or wet form; the scientific term for acid rain.

acidity: A measure of the amount of sourness in a given substance.

acid rain: The commonly used term for wet acid deposition.

acid shock: A sudden, intense addition of acid to a lake or stream, caused by the melting of acid ice and snow; occurs at the time of the spring thaw.

alkaline: The opposite of acidic; having a pH of more than 7.0; baking soda is an alkaline substance.

aluminum: A common metal on earth; in its bonded form, aluminum is harmless; however, when it is "unglued" from the soil, it is dangerous to many animals, including humans.

bonded: A term meaning "glued" or "stuck" in a chemical way; when aluminum and other metals are bonded to rocks and soil, they are not harmful to the environment.

buffer: A material, such as lime or baking soda, that can chemically weaken acid so that it is less harmful to the environment.

built environment: Those parts of our world that have been created or designed by people.

chemical reaction: A change that occurs in a substance when acted upon by another substance; sulfur dioxide, when it comes in contact with oxygen and water droplets in clouds, changes to sulfuric acid. This change is called a chemical reaction.

"dead" lake: (Also called "acid-dead lake") a lake that has become so acidic that fish, insects, birds, and other wildlife cannot survive in it.

decomposition: The natural process of decay in the environment.

dirty coal: Coal that is high in sulfur; when burned, gives off a lot of sulfur dioxide.

dry deposition: Acid pollution that falls to earth as particles of soot or smoke.

ecosystem: A community of living organisms that exists within a certain territory.

fossil fuels: Coal, oil, and natural gas, formed from dead plants and animals thousands of years ago.

guard cells: Parts of a leaf that allow tiny pores to open when sunshine or moisture is needed.

liming: A process whereby lime is added to acidic water to buffer it.

logarithmic: A scale or formula for determining magnitude that works in multiples of ten.

mercury: A metal usually bound to rocks and soil; when it is released, mercury can be fatal to humans.

nitric acid: An acid formed from the pollutant nitrogen oxide; one of the ingredients of acid rain.

nitrogen oxide: One of the pollutants that cause acid rain; produced by gas-burning engines, such as those in trucks and automobiles.

pH scale: A scale used to measure the amount of acidity in water or other substances; pH means "potential for hydrogen."

scrubber: A special filter used to remove much of the sulfur dioxide from a power plant smokestack; other, smaller scrubbers are installed in air-conditioning systems to filter out pollutants that may harm the contents of libraries or museums.

sludge: A gooey mixture of lime, water, and sulfur dioxide that is left in a scrubber.

smelting plant: A place where metals are removed from ore; these plants pump a great deal of acid rain-forming pollution into the air.

sphagnum moss: A thick mat of algae that grows in acidic water.

sulfur dioxide: One of the pollutants in acid rain; sulfur dioxide is produced by burning coal.

sulfuric acid: One of the acids in acid rain; sulfuric acid is formed from sulfur dioxide.

superstack: A very tall (between 500 feet and 1,300 feet) smokestack on a power plant; it was once thought that such tall stacks would eliminate air pollution by sending the poisons high into the atmosphere.

Waldsterben: A German word meaning "forest death."

wet deposition: Acid pollution that falls to the ground mixed with precipitation.

Organizations to Contact

The following organizations are concerned with the issues covered in this book. All of them have publications or information available for interested readers.

The Acid Rain Foundation, Inc.
1410 Varsity Dr.
Raleigh, NC 27606

Acid Rain Information Clearinghouse
33 S. Washington St.
Rochester, NY 14608

Environment Canada
Environment Protection Service
25 St. Clair Ave. E., Sixth Floor
Toronto, Ontario, M4T 1M2 Canada

National Academy of Sciences
2101 Constitution Ave. NW
Washington, DC 20418

Suggestions for Further Reading

Robert H. Boyle and Alexander R. Boyle, *Acid Rain*. New York: Nick Lyons Books, 1983.

Betsy Carpenter, "Yes, They Mind If We Smoke," *U.S. News & World Report*, July 25, 1988.

Kathlyn Gay, *Acid Rain*. New York: Franklin Watts, 1983.

Denise Grady, "Something Fishy About Acid Rain," *Time*, May 9, 1988.

Jon R. Luoma, "The Human Cost of Acid Rain," *Audubon*, July 1988.

Jon R. Luoma, *Troubled Skies, Troubled Waters*. New York: Viking Press, 1984.

Christina G. Miller and Louise A. Berry, *Acid Rain*. New York: Julian Messner, 1986.

John McCormick, *Acid Rain*. New York: Gloucester Press, 1986.

Thomas Pawlick, *A Killing Rain*. San Francisco: Sierra Club, 1984.

Index

Picture Credits

Photos supplied by Dixon & Turner Research Associates, Bethesda, Maryland

Cover Photo: J. Zimmerman/FPG
AP/Wide World Photos, 14, 19, 30, 56, 59, 66, 69, 71, 79
British Information Services, 28
© Bo Brown, reprinted with permission, 62
Robert Caldwell/Albion Books, 21, 63, 73, 74, 80
Canadian Department of Fisheries and Oceans, 40, 41
Environment Canada, 24, 31
German Information Center, 49, 76
InfoGraphics, 39
Library of Congress, 18, 26, 72
Jim Mackenzie, World Resources Institute, 46, 51
© McClain Photography, 1988, 34, 43
The National Acid Precipitation Assessment Program, 22
Mike Peters, © 1984, United Features Syndicate, 32
Pfizer, Inc., 82
© 1981, John Trever, *Albuquerque Journal*, 60
U.S. Department of Agriculture, 50, 53
U.S. Department of Energy, 84
Scott Willis/*San Jose Mercury News*. Reprinted with permission, 17

About the Author

Gail Stewart received her undergraduate degree from Gustavus Adolphus College in St. Peter, Minnesota. She did her graduate work in English, linguistics, and curriculum study at the College of St. Thomas and the University of Minnesota. Ms. Stewart taught English and reading for more than ten years.

She has written forty books for young people, including a six-part series called *Living Spaces*. This is her first Overview book.

Ms. Stewart and her husband live in Minneapolis with their three sons, two dogs, and a cat. She enjoys reading (especially children's books) and playing tennis.